Universal Truth Through
Simple Observation

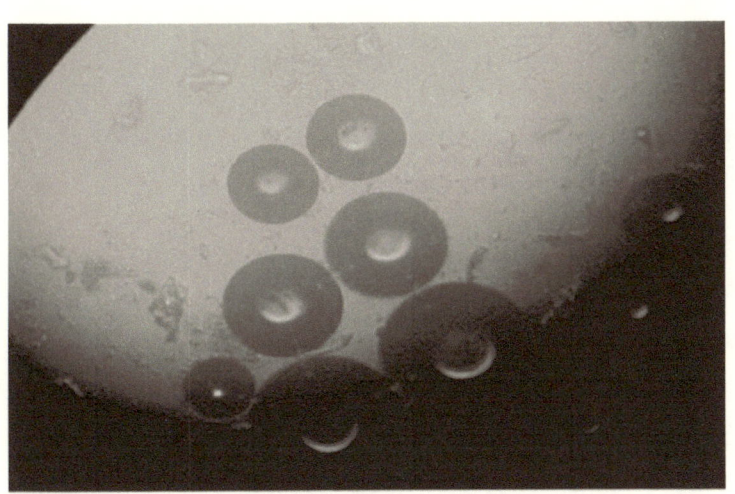

Antonius Jones

ISBN: 978-1-4116-1583-0

Published by Lulu Publishing
www.lulu.com

www.lulu.com/antonius

The Electro-Photonic Universe And A Valid Beginning

Let us consider ourselves in a vast system of oscillating particles (energy) which vibrates freely across our universe. In fact every physical object (matter) is nothing more than the invisible accumulation of sub-atomic particles vibrating at super high speeds, which in essence maintain the physical object you are observing on a physical level. Well I'll give a simple explanation: all energy is relative or related in some way and everything in our physical universe is nothing more than vibrations of unseen quantum energy spanning in a sort of micro-wave. We are actually in a huge, ever expanding electro-magnetic "soup" coupled with photonic energy, which store and transfer encoded information. Where does this photonic energy come into the equation? Atoms (the building blocks of all things) have a certain intelligence stored into their core construct and this information is exchanged by process of (coined term) electro-photonics. The electric-atomic properties of an atom need a means of transport and a proper medium along with a way to encode all the available information thus you have photonics. Light is essential in the physical universe, the evidence of this relationship (electricity and light) is right in front of your face. In any CD player what do we have? A disk stored with various information but how do we obtain it… by the use of a particular photonic oscillation, a laser. The CD player focuses the laser beam on the track of bumps electronically encoded in the CD itself. The laser beam passes through the polycarbonate layer, reflects off the aluminum layer and finally hits an opto-electronic device that detects changes in light. Photonic energy and electricity are relative and share a special bond.

Now how does all of this effect me? This is the big question, well electro-photonics effects us all in virtually every level of our lives. Human consciousness is electro-photonic as well as all of nature itself being that information generation and exchange is a constant in our universe. Even bacteria has intelligence to some degree encoded, so where is this information stored one may ask. Think of the universe as one huge super computer which is "online" and essentially connected to the same server, well it's known that everything in the physical universe can be broken down into "electro-units" (all matter consist of atoms, and all atoms must contain electrons) and these units which oscillate extremely fast COMMUNICATE with one another via photonics. Information is actually stored in each of these electrons that compose the entire universe, thus the universe cannot be "by chance" but rather a specific design but I will talk about that later. The evidence is in our face at all times but it is up to us to come to conclusions and realize what is only logical plus makes perfect sense, this is why I'm writing this very book and the conclusions I have come to. The universe is a "living network" of information available through the basic units of electron and photonic energy interface. As science tells us, everything in the physical world can be broken down into invisible particles which are known as atoms. Atoms are then composed of even smaller particles known as electrons which move at super high speeds within the atom itself. We can conclude that within the atom itself electrons in fact move so fast they create a type a friction which in turn maintains the atoms physical structure.

It is less known that a single electron has a huge storage space capable of storing massive encoded information. Now lets look at the human brain which has an almost unlimited storage potential, actually due to "electrical relay" and electron communications on more that one level. All matter is united and compatible on a sub-atomic level, just as all matter can be broken down in to sub units of electrical osculation and communication. In fact the living universe is teeming with unseen and encoded information by means of these electron units and there inner composition. In fact it would be almost preposterous to say anything in the universe is at rest, because every particle and their sub-atomic unit are vibrating constantly. This movement of sub-atomic particles creates <u>energy</u> on a more visible, physical level if that makes sense. The key to understanding the "electro-photonic" process, is <u>understanding the communication of these electrons on a quantum physical level.</u> Examples of this type of communication are radio waves, radio waves can be broadcasted very simply and easily. What is the main factor in this information exchange? That's right electrons, the communication between electrons is the key.

Photonics is key in understanding exactly how electrons communicate and understanding all energy in all forms. Photons can actually stimulate electrons and vise versa, the photon is a sub unit of the electron and all it's functions. A photon is merely an electromagnetic wave in space time, in fact we as humans actually see energy, light is the determining factor in our vision, the ability to see matter by way of "photonic dispersion". If our minds work by way of electro communication on a sub-atomic level every thought generated by the physical human brain is just electron communication to some degree. Beyond even the physical human brain unit is human consciousness (consciousness cannot be solely a function of our physical brain alone)which also is encoded by way of electron "interface". Many of these ideas may seem rather "far fetched", but I am sure this is the way the entire universe functions, *as one huge network of electro-photonic communicating individual electron units working together*. When light hits an object what happens to the energy of the light wave depends on the natural frequency in which electrons vibrate within an object. Light may be absorbed, reflected, scatted or even pass through a material. What does this tell us as intelligent observers? That light communicates directly with electron vibrations, variable to the electronic frequency of the object/material. Color is just light vibrating at a certain frequency , it is actually the photons "reading" the electrical frequency of a material which gives us visible color in which we see. All in all can we conclude that photonic energy can "read" electron activity, by means of certain electrical frequencies emitted naturally by an object. There is **a relationship between electricity and light energy** that is usable and functional, and thus we have electro-photonics.

The electron web

The electron web is a term in which I created to try to illustrate the way in which electro particles behave and interact on a quantum level as well as an observable level. If one can see the universe as one huge, lets say spider web. Every individual particle in this web represents an electron unit which composes the web as a whole. The individual web strands (made of many web particles) <u>bond</u> together to create a extremely strong construct. All units of the web are connected and all counter parts are sticky, the webs adhesive construct illustrates well how <u>"electrons will cling to other electrons"</u> . The web will and will always have a center of which it was spun and created from. Anything that moves on or across this web generates a movement of the entire web, a disturbance if you will. Let's name the spider space/time, the spider is use to moving across this web freely and undisturbed but this creates movement on the web grid and every strand on the web vibrates and adjust to the spiders every move.

This is the webs resonance, to adjust to any free movement or vibration created on the grid. Well I don't know how well "the electron web" analogy worked to illustrate this idea or point. The point of the entire analogy above is to show that spanning from a single point all energy is relative, vibrating and can be broken down into <u>communicating</u> electrical units. **All energy is relative and related on a quantum physical level, and it is this quantum particle relationship that composes the physical thing we see all around as well as the elements we cannot see.**

Electrons in communication have the ability to store or hold vast amounts of encoded information, it is by way of constant electron communication that all information is stored and accessed by these communicating units. Photonic energy would not be a working or functional process without these communicating electrical sub-units. Light has the ability to store information but only by way of the communicating electron units, which are in constant communication with one another. Electricity has the ability to flow "through" an object only by electrons reacting to other electrons within that object, this is in fact a form of communication in and between electron units. Electrons, and the flowing of these electrons (electricity) is indeed the fundamental property of all matter in the universe.

Science does explain to various degrees what electricity is and how it works, but however what makes the flowing of electrons so "universal" and usable in nature. Well, electrons are able to relate to each other via there **oscillation speed as well as the frequency of vibration patterns**. It is commonly known that electrons traveling at very high speed and in some cases can travel faster that the speed of light itself, we can even measure electron speeds in nanoseconds. These fast moving electrons are in constant motion which in turn implies that everything, all matter in the physical universe is actually in motion, vibrations of these electrons are which all atoms are derived from. The speed of these electrons is key in attaining the very structure of the atom itself, and any constant high speed is necessary to maintain a solid structure of an atom, linked to other atoms which make an entire physical object.

Understanding our universe and a true science

In order to understand any of the above and how this information effects every aspect of your life one must have an accurate view of nature in turn understanding a true science. *In order to understand anything fully and completely one must abandon all personal views and disbeliefs by having an open mind, but at the same time use a level of criticism and personal observation to come to a conclusion.* It is explained to us by science, that all material bodies are made or composed of matter and that matter is composed of unifying particles known as atoms. The physical material and matter itself is manifested in nature all around us, in simple everyday elements such as water, air and so on. How about mental objects in nature and our natural world to what degree does science explain our mental objects such as thought. Everything in the universe is composed of energy and to a greater extent vibrations, and the oscillation of particles. **Present day sciences do not explain well enough, these so-called abstract models of our universe such as thought and consciousness**. If everything is in essence energy then why is it so hard to imagine our very conscious existence and thoughts are also a frequency of oscillating vibration or energy related.

When speaking about the physical human brain, our very senses are electrical signals interpreted by this complex super computer, our brain. In essence our physical experiences are the consequences of our consciousness, so is consciousness the only reality. With *our reality being continuously shaped by our thoughts and conscious existence* is it safe to say that no solid reality actually exist but is rather an interpretation of our own thought based consciousness. How do what we actually see, hear, touch, taste and smell shape our individual experience, these functions are internal representations of an external world. Can we trust our internal representations of the "outside" world with total absolute accuracy?

Science should begin with such important questions and be able to answer questions that are relative to every single individual on every level. Consciousness should be the foundation of our sciences, because it is the stepping stone of which all individuals view reality and is necessary to understanding all scientific truth. It is by understanding our own consciousness that we begin to understand the entire universe and start to understand true sciences as well as nature itself. Maybe sciences are not ready to take on such huge responsibility, such as the development of intellect to reach universal truth because it is much easier to remain ignorant and stupid. Matter and mind are the two main elements in our universe, and matter is only relative because our minds define matter by our experiences. Development of the intellect via consciousness experience is indeed the purpose of the entire universe, to understand the self and expand our understanding. In fact what is material reality?

could material reality just be a test of intellectual understanding and for expanding our individual consciousness. I now understand the meaning of light and it's purpose in our universe, to maintain and transfer information from source to source in an on going exchange. This would make perfect sense as to why in nearly all near death experiences the individual experiences some sort of light impression. Light has the ability to retain and transfer massive amounts of encoded information, just as explained earlier in this text. *The more one understands about science the more one is pointed in the direction of the realm of consciousness to understanding the self.* It seems that all triumphs and failures stem from the self and our own personal thoughts and experiences. Why is it that the majority of people do not question, not only science but everything which points to "no where", **it is by this blind acceptance that no change will ever come in the realm of science or other wise.** The focus of today's sciences are mainly the pursuit of vast amounts of money and the development of weapons and other destructive materials. How can any "true sciences" be taught or learnt if mankind is pre-occupied with matters of such trivial nature?

The Center Of It All

How does a plant, human or animal grow to become a full functionally living "unit"? from a single point, a single bit of genetic information a blueprint if you will that physically makes you who you are. From this deduction we can come to the conclusion that the entire universe is based upon the rules of singularity and that mostly everything in our universe have a "point of origin". The evidence is all around us, if we have a mind to look. Planets also have a point of origin, the core is in fact the backbone of a planet and it is the center point of that mass. **It seems no matter where we look, we see evidence of a center point in all of nature.** If we look at an apple we notice it's outer beauty but however without the "encoded information" within an apple's seed, the apple would not exist. The most amazing part is that all the apple's physical attributes and it's physical appearance is governed by this information, or the "instructions" within the seed itself, a single seed and the result is a single apple.

Universal truth is all around us we just have to have an eye for it, it seems the more we observe nature and it's functions the more we understand about the way the universe functions as well as ourselves. The big bang theory is accurate in many ways, but sciences have neglected major questions of such a happening. I really don't want to get into "proving" anything to anyone, I just urge people to come to their own conclusions based upon observation, study and common sense. Science does not give any meaning or value to the universe, and neglect a functioning, usable universe by considering the entire universe as a random happening. My observation is quite different, I see a universe that is extremely functional and by no means a "random bang" but rather an "intelligent bang" with a purpose. Can we conclude that most of nature around us is usable and has function? Could there be a design in the universe? I say indeed there is a design, an intelligent design to our universe, but thus we encounter a problem.

Intelligence MUST be needed to create any design or give function to anything, correct? To my knowledge science has only began to touch upon the surface of any real understanding of our universe and/or it's functions. It seems to me that **science is obsessed with finding formulas and creating "models" of how the universe functions**, and this could be dangerous because it blinds us from the simple truth of our universe. Yes, numbers and math are important to a degree, but never the less they are languages that are expressed to illustrate an idea. What if we are missing factors in many of our equations and models, then does our universe fall apart? Before we can use any languages, even before models **we MUST understand basic universal law and the function of our universe or all of our equations are inaccurate**. As established above the entire universe is indeed intelligently designed and "crafted", and to any observer this in itself states that intelligence did exist before creation, before any "bang". what comes first if we look at nature, the apple or the seed in which it is from? After all we, as humans begin as merely a single cell but even before this single cell there was intelligence in the form of a genetic code.

Intelligence is found everywhere in nature, governing all interactions we encounter within nature all around us. **To understand these basic laws and ideas is to understand the foundations of any true sciences**, then and only then can we advance to the point of absolute scientific understanding. In nature there is always a "center point" , rather it is the nucleus of each and every one of your bodies cells to the core of a planet, there will always be a center and this is fundamental truth number one. The second fundamental truth found in nature is that *everything spawns or spirals from that center and this "center point" is unchanging and governs every action or process outside of itself*. The third and final fundamental truth is that *there exist intelligence in all of nature and it's functions,* intelligence also exist before any physical manifestation in nature. Is it possible to build anything at all without having an idea or plan behind it? In the universe there is a plan behind every bit of matter seen and unseen, a blueprint for every single "unit" in our universe. What is the "core" or center of our being, our core construct as human beings?

Consciousness is in fact the center of us all, and governs all our experiences internal and external. Consciousness should be one of the main focuses of our sciences, that is if we ever hope to understand the entire universe. How can we expect or even what to understand the universe without understanding ourselves? In my mind this creates a huge problem in sciences and in a society, I think it is time to focus on what is important at this point in history.

An Intelligent Evolution?

 I understand that the subject of evolution has much controversy surrounding it, but it is necessary to touch upon the subject and address a few things. The material here is a result of much study on the subject and my own personal observations along with our present day common scientific knowledge to form a solid fundamental idea. Evolution is a change in a species, a species is nothing more than a product of it's own environment and the adaptation of that environment. Darwin explains that a species will adapt to it's environment and over a period of time, change in physical characteristics to ensure it's survival. Is adaptation, not a form of learning? Could we say that any change over any period of time is due to direct learning from our "outer" environment or habitat? Darwin explains various animals that adapt and become smarter and more adaptive animals, such as a bird which develops thicker feathers over time to survive in a less temperate climate.

A more adaptive bird is the product of direct learning in an environment, a bird becomes a "smarter" and more adaptive bird in this case. How often do we hear of a species developing into a totally new species? let's say a fish that becomes an adapted monkey, does this change or adaptation sound rational, why does our sciences tell us that such changes are accurate? A species has the in-built ability to learn from it's environment, even bacteria has the ability to learn from it's immediate environment. It seems that nature is much more intelligent than we give it credit for and gives us many tools of development. Any change of a species (including humans) must be due to understanding and conscious development. *Adapting is just a form of learning* and we can conclude, through logical deduction, that learning involves consciousness and intellect. Does an external environment help us develop our internal environment, our consciousness?

In the case of an evolution, would a species of monkey or apes evolve into present day man? Evolution is a fact and does occur, but evolution is not responsible for any species that changes into another species even over time. Do we see evidence of fish changing into any species of bird, even over a period of time? A fish in some instances may develop (through mutation or by other means) a type of wing or "wings" like in the case of a bird, but in any pretence that fish remains a fish in that species category. Science should not accept such ideas fully but should look at all possible angles of an "evolution" and it's effects on a species.

The only evolution, is the evolution of the intellect through direct physical, environmental experiences in which a species will experience. This is the only "evolution" that makes sense, an evolution through adaptation and by learning to adapt to an environment a certain intelligence is needed to make such changes. We as well as the animal kingdom learn through a set of physical experiences to come to any change, a certain amount of knowledge is needed. **Can we conclude that the entire universe is merely a playground in which all species can develop and learn?** I can not stress enough the importance of the "conscious universe" and the realm of the human mind, then and only then will we begin to understand the secrets of the entire universe. In the case of man, a man will only become a more adaptive and environmentally adapted man.

A human will ONLY become a more adaptive and smarter human being over time, a human will never "transform" into another species even over a period of time. In any evolution the environment in which a species inhabits will always a major factor in the process of an evolution. A monkey or an ape will never evolve over time into a human being but only evolve into a smarter more adaptive monkey or ape. A human in a habitat may over time adapt to an environment and develop a physical appearance similar to an ape, but nevertheless will just be an "adapted" human. Evolution does in fact take place, but an evolution will not include the transformation of a species into another species of a different "category".

A Simple Truth

Is there such thing as a simple truth, a set of ideas that can help people live better, simpler and a happier lives? Indeed there is a simple fundamental truth, and that truth lies within the very roots of your own conscious mind and truth will be brought about by simply understanding the universe around you but mainly the universe within you. The more you understand about yourself, the more you will understand others in turn the more you understand nature around you the better you co-exist within it. What if no one wanted to understand or learn anything?

At this time in human history in our present day it is EXTREAMLY important to understand nature around us and how we are integrated within it, after all we must co-exist in nature. My own personal observations obligate me to illustrate the immediate dangers of pollution in our environment and in nature. Have you noticed through personal observation the increased frequency of natural disasters? Have you noticed the constant and dramatic increase and decrease of temperature all over the world? This is a perfect example of a "simple truth", are we, as individuals on this planet able to observe simple and important happenings around us?

Pollution is a major problem in our present day society, and will continue to be so in the immediate future and if nothing is done about this problem it could then effect us all in the form of a GLOBAL WARMING . Yes, we have all heard of the problem of Global Warming but how much do we actually know about the problem at hand? Through my own personal observation and extensive study I have found that this issue of a Global Warming have many underlying effect and severely effect every ecosystem on the planet as well as every individual on the planet. I will not go into this issue extensively but **I just wanted to illustrate the value of finding a simple truth through observation and understanding** and how important critical think is in our universe. To find any truth we must deeply evaluate our own conscious thought and understand our own inner workings, and **a simple truth does exist but will only be brought about by a conscious effort to understand nature and ourselves.**

If we never have the desire to want to understand anything about ourselves or nature then important issues such as the unseen effects of Global Warming will go unnoticed and this is dangerous for us all. Individual intellectual development and growth must take place within each and every one of us, and that is the simple truth.

The dangers of a materialistic society

What is more dangerous than the exploding of an atomic bomb in your very hands? Some may say what could possibly be more dangerous and harmful than an atomic bomb going off in your hands, well the answer may surprise you. Living in a materialistic society is far more dangerous than any physical catastrophe that can be done to us. Is it not obvious that the entire universe is based on the principals of adaptation to an "environment", adaptation to any environment requires direct use of intellect and results in learning. In fact intellect is an abstract function in all of nature, even though man may not act as it's counterpart the fact remains that we are all a part of nature and it's processes.In a totally materialistic society is there and adequate release for the eternal human psyche, a way to truly explore ourselves and natural processes?

Accumulating vast amounts of material items or goods is actually very dangerous to our conscious development because it shifts our focus off any internal development of the psyche to material matters. Have you ever noticed, no matter how much or many material items you accumulate you are never feel "complete" inside. A materialistic society will collapse on itself because it's foundations are weak, the entire material universe was created for us to learn, by the process of direct learning within an environment rather it is a class room or an entire universe. Why give each and every living being a consciousness combined with an environment this equals direct learning.

As humans with our advanced abilities of reasoning and logic we sometimes forget or never see how simple the entire universe is, a giant classroom, a university of learning. Material reality gives us a unique environment for learning, as I say "how can a child learn without other children or toys to interact with". The intellect and our very consciousness are indeed abstract, where exactly is consciousness? When we die our physical bodies will remain the same until the biological process of dying fully takes it's effect, so where is our consciousness at this point and where is it while we are alive? If you were to ask some of the world's greatest brain surgeon's such as Wilder Penfield, they would simply say that the source of consciousness is not located in the brain itself but seems to be "somewhere else". The greatest problems often originate or stem from within society itself, society is the very structure of which we live and base our lives by. Materialistic societies have been the down fall of almost every so-called great society in our past, rather domestic or afar. If we continue down this path what will become of our societies here on earth? As technologies advance and progress we become more dependant on gadgets to solve our problems and push buttons accessories.

Technology has it's ups and downs so don't get me wrong, but the fact remains without a proper and fundamental knowledge of nature and the universe as well as ourselves we our actually just hindering our conscious development. The best way to understanding the entire universe and all things around you is to understand and evaluate yourself, to understand yourself and the choices you make. The point is to develop the intellect by simple observation until you can understand the reason for almost everything you encounter which leads to an understanding of the entire universe. What can you do and why you may ask?

You can start by observing what is around you, study the choices of others as well as yourself until things start to make sense, then maybe you can begin to express your ideas with others and maybe even make a change in your environment, community, society or the entire world. It is learnt in every science text book that every object in the physical universe consist of atoms, electrons and so on and these units are nothing more than recycled energy, electromagnetic oscillations that will one day cease. Mankind's greatest danger is materialism, by adopting a materialistic attitude and way of life we neglect the development of the psyche. The only worthwhile advice I can offer all readers is to never stop thinking or questioning what is around you.

Nature the simple and perfect "program"

  The simple method of observation is a useful and fundamental tool in understanding ourselves and the universe in which we are a part of. Nature, what better physical model to observe and to learn from? Indeed, we are conscious for a reason and have other interactions with humans and nature around us for a reason. How complex is nature as a functioning "program" in our electro-photonic universe? Is nature not as beautiful as it is functional? Understanding the entire universe and ourselves as a part of this system starts with the simple observation of what is right in front of our faces, nature itself. It is up to us as intelligent human beings to put that intellect to work, to understand what is in front of our very faces. How much do we actually even know about our very own bodies and minds?

The more we understand nature and its functions, the more we in turn learn about ourselves and the entire universe. The best way to understand this "perfect program" is by direct observation, to learn by the observation of nature and all its functions. If we as humans choose to do our own will and go against nature instead of "going with its universal flow" could be the hugest mistake we could possibly make, rather on a global scale (due to our collective ignorance) or individual. Nature and all its functions exist for a reason, as a tool for every individual conscious being to "build upon" and learn from as a model for observation with the final goal of LEARNING. It could be said that the entire universe is a playground for conscious development.

Would you say a child learns better and faster with or without other children and "toys" to interact with? Would a child learn in the same way with no interaction or no toys, perhaps just darkness? The same can be said about the existence of the entire material universe, a huge playground of learning? If this does not make sense to you perhaps i am doing or explaining something wrong and maybe need to adjust my methods. I'm sure that i can not tell you something in which you are not already aware of in this universe (through your own universal collective consciousness, the universal knowledge bank which we are all connected) , I can only bring in to your conscious attention. What happens when we pollute nature's functions and ecosystems with cluttered pollution which only seems to be increasing daily on our planet. Does nature in all most any of it's functioning "programs" including the ecosystem A.K.A weather patterns have a way of cleansing itself from such a pollution?

Almost every animal, plant and even ourselves, human's have the in built "response" to undergo self-repairs when encountering sustaining damages. As humans we clot when we have an open cut, plants usually regenerate an entire stem or vine when severed so is our environment or ecosystem any different? When encountering vast pollution which is damaging to the ecosystem what happens, well what happens when we are exposed to massive clutter or harmful debris when inhaled in the atmosphere? We sneeze to expel all harmful molecules which were inhaled, so in the same sense maybe if we expose the intelligently designed ecosystems to harmful debris and pollution maybe it to will sneeze in the form of extremely violent, strong super storms and other upheavals? Nature has an in-built self correcting design by any means and the faster we actually realize this and respect this the closer we can become to recovery and prevention. The universe is built upon a freedom of choice and total freedom to do as we choose but all actions in turn have consequences in the long run but answers are always right in front of our very faces if we can LEARN BY SIMPLE OBSERVATION.

Global Warming ??
- A brief overview

How serious is this Global warming issue, and to what degree does it effect our environment and essentially our lives. We are all aware of the issue of accumulating pollution on a global and local level, but are we fully aware of the effects of pollution in our global environment? According to much personal research Global Warming has many underlining effects on the environment and most of it's effects will go unseen until manifested physically, let me explain. Are you aware that we actually live on a huge nuclear reactor? Yes, the earth is in fact nuclear in nature and that all usable energy is created within the earth itself including all heat. The bulk of every process occurs within the earth itself and manifests itself on the surface. Surface temperatures serve the purpose of maintaining the internal temperature and structure of the core of the earth itself. The recent tsunami disaster in Asia may only be the beginning of more intensive natural disasters. Have you noticed the recent increase of natural disasters all around the world? Perhaps, it takes extremely dangerous natural disasters to wake us to the reality of how miniscule we are on this planet. The question is why are there more frequent and deadly natural disasters occurring all over the world? Science tells us that global warming is indeed a "real problem" in which we face due to our pollution of the environment.

The "greenhouse" effect which many people are aware of is said to be a huge problem within our environment due to these greenhouse gases, the most common of which being Carbon dioxide. Greenhouse gases are effecting almost every atmosphere on our planet daily, it is this build up of Atmospheric greenhouse gases which is extremely dangerous. The industrial revolution brought about great change in our world. The earth's atmospheric concentrations of carbon dioxide have increased nearly 30% and methane concentrations have more than doubled as a product of the industrial revolution. Chemicals such as nitrous oxide have increased in concentration about 15%. These increases have enhanced the heat-trapping capability of the earth's atmosphere.

On a brighter side Sulfate aerosols which is an air pollutant, helps by cooling the atmosphere by reflecting light back into space but sulfates are short-lived in the atmosphere. Global warming has many visible effects on our planet, but what about what we can't see? How about the core of the earth itself, the core of our planet is in fact very important and many functions. Science should study the core of the earth itself for an answer to the global warming issue and how we may be effected, the surface of the earth is in direct correlation with the core itself, being the giant nuclear reactor that it is. The increase in natural disasters, floods and the recent tsunami in Asia may only be the beginning of up coming natural disasters that could indeed ravage our planet. Pollution levels rise almost daily on a global scale, as we squander for "oil" and other such fuels we are indeed globally, accelerating the destruction of the atmosphere and ecosystems at an alarming rate. After years of study (since grade school) and simple observation i have come to these conclusions which are indeed very real legitimate issues which need to be taken care of. Personally i have taken many steps to make these findings fully known, the most important issue in which I have tried to explain to the establishments is THE UNSEEN ISSUES of a "global warming" which in fact does not necessarily mean just the heating of the surface of our planet but also the over heating of the earth's core itself and other climate flocculation's. As it is known on our planet we experience seasons (fall, winter etc.) for a very important reason....

to maintain the very source of all the earth's energy, to maintain the earth's core temperature itself. Indeed the earth's poles, north and south are to remain cool for a reason wouldn't you say? Well what can be concluded when these poles begin to melt as they are so now? The poles have noticeably begun to melt and actually rise in temperature about three degrees and this is not a favorable event. The poles of the planet have few very important roles one of them being to maintain the earth's equilibrium and to sustain "inner" temperature, in essence the core's temperature. I ask of every person reading this material to use your very own intellect and "simple observation skills" to come to your own conclusions but do NOT sit around and do nothing about such dangerous situations. Do not act as sheep do and just blindly follow, I urge you to use your own imagination and intellect to come to this conclusion that the earth is and will continue to be in danger if NOTHING is said, if NOTHING is done and if no one STANDS UP!! My goal of this book is to stimulate minds possibly wake some up, I would love to see people start using their imagination along with a questioning, logical mind and by the power of OBSERVATION start to change the world. It is easier for groups of strong minded, intelligent people with vision to change things rather than scattered single individual strong minded, intelligent people with vision to change very non-intelligent ideas, laws and beliefs.

About The Author

My name is Antonius, I am an individual on a mission. My mission is to expand on today's sciences to come to a unifying set of ideas that are not extremely complicated but rather simple. Yes, I am twenty years old but believe it or not I have been conducting research, writing (various materials/poetry/ theories ect.) and studying an array of different topics since the age of twelve. I have studied world religions to come to a understanding of spirituality, sciences, computer systems, you name it.

I would consider myself an observer in nature and our universe, a simple observer. Now is a perfect time for me to express my findings and studies with the entire world, but I still have so much more to learn. This book was created to share my research, studies and observations on just about everything around us but mainly the structure and design of our universe. My goal is to illustrate that the universe is not as complicated as some believe it to be, but is rather simple. Nature and the universe are based on simple easy to follow universal rules. Simple observation is the key to the entire universe. At this point in human history it is time to address certain urgent issues. My book is intended to urge readers to observe the universe in a whole new light and expand their Consciousness and awareness, I ask all readers to explore "Universal Truth Through Simple Observation" with and open mind. My book is not intended to be big and bulky and just fill page after page with thoughtless clutter but rather to focus intensively on one point or subject at a time with powerful expressive statements that comprise a smaller more impacting book.

AAAhhhh to look up and see the expressions of such a genius design, proves to me that there is some intellegence in this universe even if it is not found here on earth
I then take a deep breath of relief for this is all the proof i will ever need.................

-Antonius----

www.ingramcontent.com/pod-product-compliance
Lightning Source LLC
Chambersburg PA
CBHW021932170526
45157CB00005B/2292